THE OUTDOOR ADVENTURE PLAYGROUND

GW01372454

the big melt

fire

air

the face of steel

water

earth

TO COACH DROP OFF/PICK UP POINT

MAGNA EDUCATION CENTRE

MAIN WALKWAY

PACKED LUNCH AREA AND O₂ CAFÉ

TICKET DESK

ENTRY TO ATTRACTION

THE RED HALL ENTRANCE AREA

MAGNA STORE

O₂ RESTAURANT

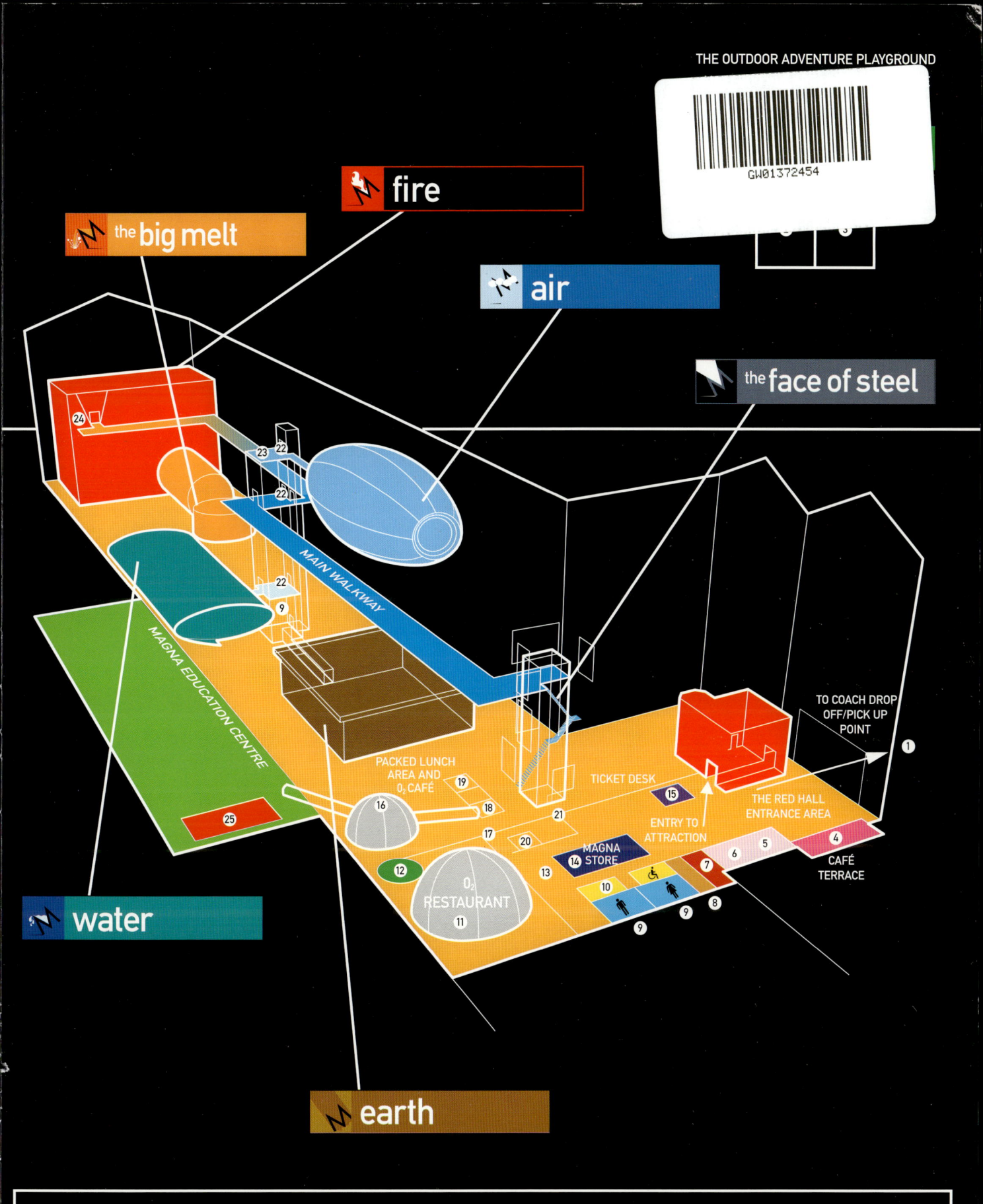

CAFÉ TERRACE

### FACILITIES

1. Coach Drop Off/Pick Up Point
2. Play – Outdoor Adventure Playground
3. Play – Food Sales
4. Magna Café (Foyer)
5. Visitor Feedback (at Information Desk)
6. Information Desk, Lockers, Buggy Park and Lost Child Point
7. First Aid
8. Public Telephones
9. Toilets, including wheelchair access toilets (on Earth level of Transformer House & in Entrance Foyer)
10. Baby Change
11. O₂ Restaurant
12. Under 5's Play Area
13. Meeting Points
14. Magna Store
15. Ticket Desk
16. Packed lunch area and O₂ Café (peak times only)
17. Entrance to Education Centre (schools use only)
18. Schools Reception
19. Schools Toilet
20. Disabled Lift
21. Lift to Main Walkway
22. Lift to All Levels (disabled access)
23. Transformer House – Food Sales (peak times only)
24. Fire Pavilion – Food Sales (peak times only)
25. Sheffield Wildlife Trust Education Centre

**EXIT VIA FIRE LEVEL**

## air

**SHORT VISIT**
Air Waves – make weird music. The Winds of Change – feel the blast of near gale force winds. Gyroscopic Chair – get in a serious spin. Air Tornado – get up close to a twister.

**LONGER VISIT**
Cloud Rings – blow big smoke rings. Body Noise – make extremely rude noises. Fly Like a Bird – experience the forces of flight. Air Cannon – bolts of air make amazing patterns. Turbulent Orbs – demonstrate how the world's weather works. Galloping Gertie – stand on a wobbly bridge while another one falls down.

## fire

**SHORT VISIT**
Electric Arc – sparks fly and steel melts. Cool Crystals – melt crystals and make psychedelic effects. Wild Fire – see the power of fire out of control. The Fire Tornado – feel the heat of a towering spiral of flame.

**LONGER VISIT**
Jacob's Ladders – sparking, flashing, spectacular art. Steel Saver – operate an electromagnetic crane. Fireworks – set colours on fire. The Story of Steel – time travelling by computer. Shaping Metals – rolling, casting, forging. The Properties of Steel – noisy interactive sculpture. Lava Lamps – their secrets explained.

## water

**SHORT VISIT**
Water World – at the entrance to the Pavilion. Mist, rain. Pouring indoors. Supersoakers – fire water cannon at human targets. Wet Play – fantastic squirty, splashy things to do with water. Hydrogen Rocket – power up a rocket and fire it into the air.

**LONGER VISIT**
The Big Blue – laser cut art. Fishy Tales – race a salmon upstream through the ages. Cool Steel – make water dance. P's and Q's – test your knowledge with the Mannequin Pis. Water Wheels – have fun with the power of water. Down the Pan – count the toilets flushing at Magna.

## earth

**SHORT VISIT**
The Diggers – operate a real JCB. Anyone Down There – tunnels to play in and explore. The Excavator Wheel – play your part in a working quarry. Danger Blasting – explode a rock face.

**LONGER VISIT**
The Earthnet – spidery art that stalks you. Role Play for kids – helmets, barrows and buckets. The Rest Hut – play your choice of story. Fantastic interactives – try them. Levers, Gears, Pulleys and Hydraulics – lifting rocks and people. Not In My Back Yard – voting on the environment, using your head.

### THE BIG VALUE MAGNA ANNUAL PASS
Adult. Child. Family. Or concessionary. The Magna Annual Pass is exceptionally good value and gives you unlimited access to Magna and Magna Play for a full 12 months. And it means you don't have to queue. If you've already bought tickets, we can upgrade them to Annual Passes on the day of your visit and deduct the amount you've already paid.

### MAGNA PLAY
Admittance to Magna Play is included in the price of your Magna ticket.

### MAGNA FACILITIES
### THE MAGNA STORE
The Magna Store sells toys, gifts, souvenirs and books, as well as a range of stainless steel gifts and homeware. Prices range from pocket money upwards and are all good value. Mail order is available, and school groups can pre-order gift packs. Call 01709 720002.

### THE $O_2$ RESTAURANT AND THE MAGNA CAFÉ
Britain's first ever inflatable restaurant serves home-cooked meals, snacks, cakes, hot and cold drinks and other goodies, and it's fully licenced. There's also the Magna Café in the Red Hall, and on weekends and holidays there's the $O_2$ Café, carts serving tea, coffee, cakes and more by the Air and Fire Pavilions. The Magna Store, $O_2$ Restaurant and the Magna Café are open to everyone. You're welcome with or without a ticket.

### CHILDREN AND BABIES
Everything at Magna is fun and safe for children. Magna Play has a special under-5s zone. There's a softplay area alongside the $O_2$ Restaurant, and special small-child-friendly activities in the Water and Earth Pavilions. Please tell your children to go to the Information Desk should they get lost – and show them where it is. Off the Red Hall, there are baby change and baby feeding facilities.

### LOTS OF LOOS
There are loos in the Red Hall, the Water and Earth Pavilions. In the Red Hall and Earth, there are toilet facilities for disabled people, with wheelchair access.

### ACCESS, WHEELCHAIRS AND LIFTS
Magna is designed for everyone to enjoy and many of the activities and exhibits are suitable for visitors with disabilities. All areas are fully accessible to visitors who use wheelchairs. There are wide access lifts to all levels in the Red Hall, by 'The Face of Steel' show and in the Transformer House. Wheelchairs are available free from the Information Desk. And one-to-one carers are admitted free.

### MAGNA SAFETY
### PLEASE BE CAREFUL
Magna has been designed with safety in mind. But as with any visitor attraction, we would ask you to be careful. Watch your step and do not let children run about, climb on or lean over walkway railings. All exhibits are pacemaker safe. Strobe lights and pyrotechnics are safe – though they may well make you jump! There are qualified first-aiders on the staff.

### MAGNA INFORMATION
### CORPORATE AND PRIVATE HIRE
### EDUCATIONAL AND GROUP VISITS
The Red Hall, Pavilions, $O_2$ Restaurant and other facilities are available for corporate and private hire. Special facilities and discounts are also available for pre-booked visits by schools and groups of 12 or more. For details, call 01709 720002. Or see inside back cover.

### OPENING HOURS
Magna is open 10am – 5pm daily. Last tickets are sold at 4.30pm. Magna is closed Christmas Eve and Christmas Day.

### LET US KNOW WHAT YOU THINK
At Magna, we welcome feedback from visitors. So please let us know what you think. Make your comments at the Information Desk or access the Info Page on our website – www.magnatrust.org.uk Thank you.

Have a good time!

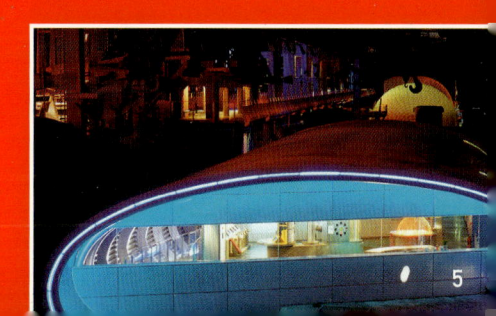

# Planning a visit

**Prepare to be shocked and amazed.** Dress warm. Come early. Or late. Hurry. Or take your time. See the must-sees. See the shows. The kids' favourites. Have fun. Have a rest. Eat and drink. Have some more fun. Come again. And again.

### NO LIMIT TO YOUR VISITS
Here are some straightforward tips about making the most of Magna. It's worth just taking a minute or two to read them. They're helpful and they'll add to your fun.

### BE PREPARED FOR ADVENTURE
Magna is an amazing experience, and part of what makes it amazing is the simple, breathtaking size and power of the place. So be prepared. It's safe. But it's very noisy. Dark and shadowy. There are bangs. Explosions. Pyrotechnics. Strobe lights and sparks. Kids love it. But stay close to little ones and each other – if you're nervous. It can also get quite cold. So in winter and on all but the warmest days in summer, wear something warm. You can always leave it in our locker room if you get too hot. There's a lot of walking to do, too. With steel steps and walkways. So wear sensible shoes.

### PLANNING YOUR VISIT
Magna is hugely popular and big as it is, it does have a maximum capacity. So on busy days – weekends and holidays – unless you have a Magna Annual Pass, you may well have to queue. It's always a good idea to call first – particularly if you're travelling any distance – for information call 01709 720002 weekdays or 01709 723130 weekends

At busy times, it's best to come first thing in the morning. Or at the start of the afternoon.

### HOW MUCH TIME DO YOU NEED?
You can whizz around and see highlights in an hour. A real visit takes at least two. But to get the full Magna experience, you need half a day. The whole day. Or more. Because there's such a lot going on, lots of people make repeat and often regular visits.

If you're going to be a repeat or regular visitor the Magna Annual Pass is spectacular value.

### WHAT TO DO WHEN
Once you've got your tickets, you're welcome to come and go as you please. You go in through the big chunk of concrete and steel in the Red Hall, which was part of the original steelworks. Then up onto the main walkway via the first of the shows 'The Face of Steel'. After that, it's up to you. There's no particular order you need to go around the pavilions in. So take your pick.

Check out which pavilion is least busy – and go there first.

### THE MUST-SEES
For maximum fun and find-out value, we've recommended 'must-see' exhibits for each of the pavilions. These tend to get busy, too. For obvious reasons. Especially the Supersoakers in Water and the Diggers in Earth. Again, check them out. Go early or late, mornings or afternoons.

Or try going to the most popular exhibits at lunchtime, when other people are busy eating!

## THE SHOWS

'The Face of Steel' is the first thing you see. It's a stunning show. In both scale and the story it tells. It runs continuously. And you really should see it. Watch it on your way in. Or come back to it. Same goes for 'The Big Melt'. Running every hour weekdays. Every half hour, weekends and holidays. It's so big, noisy and spectacular, you can't miss it. And you won't want to!

## YOUNG CHILDREN

Magna is flat-out fun for everyone. But we've paid special attention to families. If you have younger children, you might want to concentrate on the Water Pavilion, where you'll find the Supersoakers, buckets, taps, boats, floats, waterproofs and squirty things. And the Earth Pavilion, with diggers, hard hats, work jackets, wheelbarrows, trucks and stories you can play in the Rest Hut. But everywhere's great for everyone, really.

## IF IN DOUBT, ASK

If you need any general advice, help or information, ask at the Information Desk in the Red Hall. If you can't find something on your way round; lose your way, or don't know how to work something, then look out for the Magna Enablers. They're the ones in the red Magna shirts or black Magna fleeces. They'll be glad to help.

If any exhibits don't seem to be working, please tell the Enablers. We have teams of technicians on duty all the time, and unless broken exhibits need special parts, we'll get them working again, same day.

## SPLITTING YOUR VISIT

Magna is big. It can be tiring. Especially for little ones. So we recommend splitting your visit. Do two pavilions. Then retire to the $O_2$ Restaurant or the softplay area for kids. Or grab a coffee. Have a rest. Then carry on.

## ENJOY THE BUILDING

A lot of people have said to us that it's worth a trip to Magna, just to see the building. Pretty well all of the outside – and much of the inside – is original. Look up to the roofspace and the walls. Down below the walkways. You'll see parts of the old steel plant still there – intact. Hooks. Chains and pulleys. Great big vessels. These are fantastic 'interior landscapes'. And designed into this magnificent setting, the pavilions themselves are something to see!

## WHAT YOU CAN'T DO

There are very few rules at Magna. No smoking anywhere in the building. Walk. Don't run. Use the stairs if you can, so the lifts are kept free for people who really need them. Please treat the exhibits and other people with respect.

## WHAT YOU CAN DO

Anything you like. Within reason. Photography – still, flash and video – is fine. Yes, you can touch. Play with the equipment. Generally mess about and get in and amongst it. Magna is here to be enjoyed.

## WHAT YOU SHOULD DO

Have lots of fun. Find out lots of things. Have a really, really good time. And come back and see us again.

## WHAT'S NEXT

Turn the page and the Magna Souvenir Guide will provide you with a guided tour in print. Use it on your visits as a guide to the major exhibits and shows. Or as something to remember us by. There's also a brief history of steelmaking, the steelworks and Magna. Some suggestions for other places to visit in the area. And contacts for further information. We hope you find it useful – and enjoy it. Thanks for coming to Magna.

Don't forget to stop off at the Magna Store before you leave. And if you haven't done so already, ask about Magna Annual Passes. Hope to see you again soon.

**This is a flat-out, full tilt, fun, fantastic experience.** In a vast building. Surrounded by steel. Noise. Light. Fire. Smoke and shadows. This is where you find out how to fly. Make steel. Detonate explosions. Launch a rocket. Come face to flame with a firewall. Use real earth-moving machinery. Spin like a whirlwind. Splat other visitors. This is Magna. The UK's first Science Adventure Centre. No limits.

### ONE BIG ADVENTURE
Based in a former steelworks but dedicated to the future, Magna takes you on a stunning journey of adventure through science and technology. Up into the air. Through fire and water. Down into the earth. It's where you discover how much fun it can be to find out about the world around you. Getting into it. Playing with it. Experiencing it. Making a noise about it. Just plain enjoying it.

### MAGNA MEANS GREAT
Magna is huge. The building is over a quarter of a mile long. 42 metres high at its highest point. Once, it was the melting shop of Europe's greatest steelworks. Now it's come to life again as Britain's first Science Adventure Centre. It's a massive, magical space. With a real sense of place, and of the people and the industry it celebrates. It has amazing architecture. Great facilities. Incredible exhibits. It will take you over. Overwhelm you with fun and adventure. On a scale you've never, ever seen before.

Magna is Latin for big or great.
A great big adventure.

### AIR FIRE WATER EARTH
At the centre of the Magna experience are 4 spectacular pavilions, representing the 4 elements. Air, Fire, Water and Earth. Air is like an airship suspended in space. Fire is a black box wreathed in lightning. Water is a stainless steel wave. Earth, a great slab you tunnel under – deep beneath the ground.

The Greek philosopher Aristotle, 384 to 322 BC, first thought up the idea of the four elements. And the four elements all play their part in the steelmaking process. It needs Earth, Air, Fire and Water. Put them all together and it's quite an experience.

## NOT FOR PROFIT MAGNA

Magna is run by a non profit making charity, The Magna Trust, charity number 1074578. All income from ticket sales, retail and catering is used to run Magna, to fund developments for the future and to improve services for our customers.

The Magna project has been made possible by funding from both public and private sources:
The Millennium Commission
Yorkshire Forward
Government Office Yorkshire and Humberside
Stadium Group
Rotherham Metropolitan Borough Council
Rotherham Chamber TEC
Lombard Corporate Finance

The exhibitions and building described in this book are the first stage of the Magna project. We have successfully created exhibitions and environments where science comes alive. The emphasis is very much on having fun, but there is a strong scientific underpinning to everything we have done. We are continuing to add information and backup material for the exhibits.

## FROM MAGNA'S CHIEF EXECUTIVE...

Our aim, at Magna is to entertain, inspire and educate. We've chosen to do that with two subjects that are often seen as dull or uninteresting; science and technology. We hope you have a brilliant time finding out that isn't true. And we are just at the start of the creative process – we will adapt and change our exhibitions and programmes as we learn better what interests you and what you want to find out about.

A key part of Magna's appeal is that it is based on a strong underlying structure. We started with the Aristotelian elements, earth, air, fire and water because they fitted the background subject of Magna, steel, and helped us develop themed interactive environments. These 'elements' all play a key part in steel making and we've also been able to interpret them much more widely. Allied to the four elements is another structure; principles, innovation, application and production. We have used these to look at a huge variety of things like bridge design, mechanical diggers or a vacuum cleaner. The exhibits have deliberately been designed to show how teams of people and individuals have used innovation and invention to put scientific principles into play to make products. We have also shown this process at work in the making of art. Always our desire has been to show creativity in action. Creativity is the well spring of Magna. We can't solve tomorrow's problems with yesterday's solutions; a key purpose for Magna is helping to release and sustain creativity, especially in young people. This focus on invention and discovery requires new methods and Magna is a place where learning can take place informally. It can form the basis of a great day out or the foundation for further learning by children in the classroom.

We believe, that the creation of informal discovery environments, like Magna, backed up by more formal leaning in school is the learning path for the future. Because of their critical importance, we have applied the method to science and technology.

Our most successful exhibits defy classification as works of art or science – they are both. They can be experienced as intriguing or fascinating and they can be explained scientifically or technically. This, in part, sets our future direction. Whether we are exploring robots or product design, we will create exhibits and run programmes that make links between different disciplines. Flexibility of mind and creativity in application – these are the keys to a sustainable future.

*Stephen Feber*
Stephen Feber
Chief Executive

## FROM MAGNA'S CHAIRMAN...

Great achievements do not happen by accident. They usually experience a long development period, detailed planning, numerous surprises (not all good) some doubters, strong teamwork, the juggling of many issues, raw human emotion and a clear sense of purpose. Magna is no different. Throughout this century tens of thousands of people worked together in the Templeborough works – now the Magna building – to create steel, in a tough, dangerous and physically demanding environment. Teamwork... in the Devil's Kitchen. Since 1997 hundreds of people, more than half of them local, have worked with equal energy, stretching their creative, physical and emotional resources to give new life to this once awesome steel melting shop. And today, nearly a hundred staff, each one committed to helping our visitors have a memorable experience, stimulating eyes, ears, brain and spirit – to run one of South Yorkshire's finest achievements in recent years. I have yet to meet anyone who has not been affected by the Magna experience, and I hope you will be too. Magna is a charitable foundation, and so any profit will be re-invested in the facilities to make them even better. I would like to thank all those who helped to create Magna – our funders, our skilled contractors and partners, our staff. Finally my thanks to you for having the confidence in Magna to spend your time here.

*Chris Welsh*
Chris Welsh
Chairman of the Trustees

# Face of steel

**Evocative. Thought provoking. Stunning to look at.** Start your journey through Magna by entering the lives of the people of steel. Their world. Their community. The generations. Clocking on and clocking off. The noise and the dirt. The difficulties and dangers. Wartime. The fun and friendships. Strikes and redundancy. Endings. New beginnings. Lasting achievements. A wall of light and sound.

**THE SOUND OF STEEL**
Families lived close to the works. They lived with the noise and the dirt.
"As a little child, you didn't need to count sheep. Just listen to the steelworks' hammers".
"I remember me father coming home – black-faced".

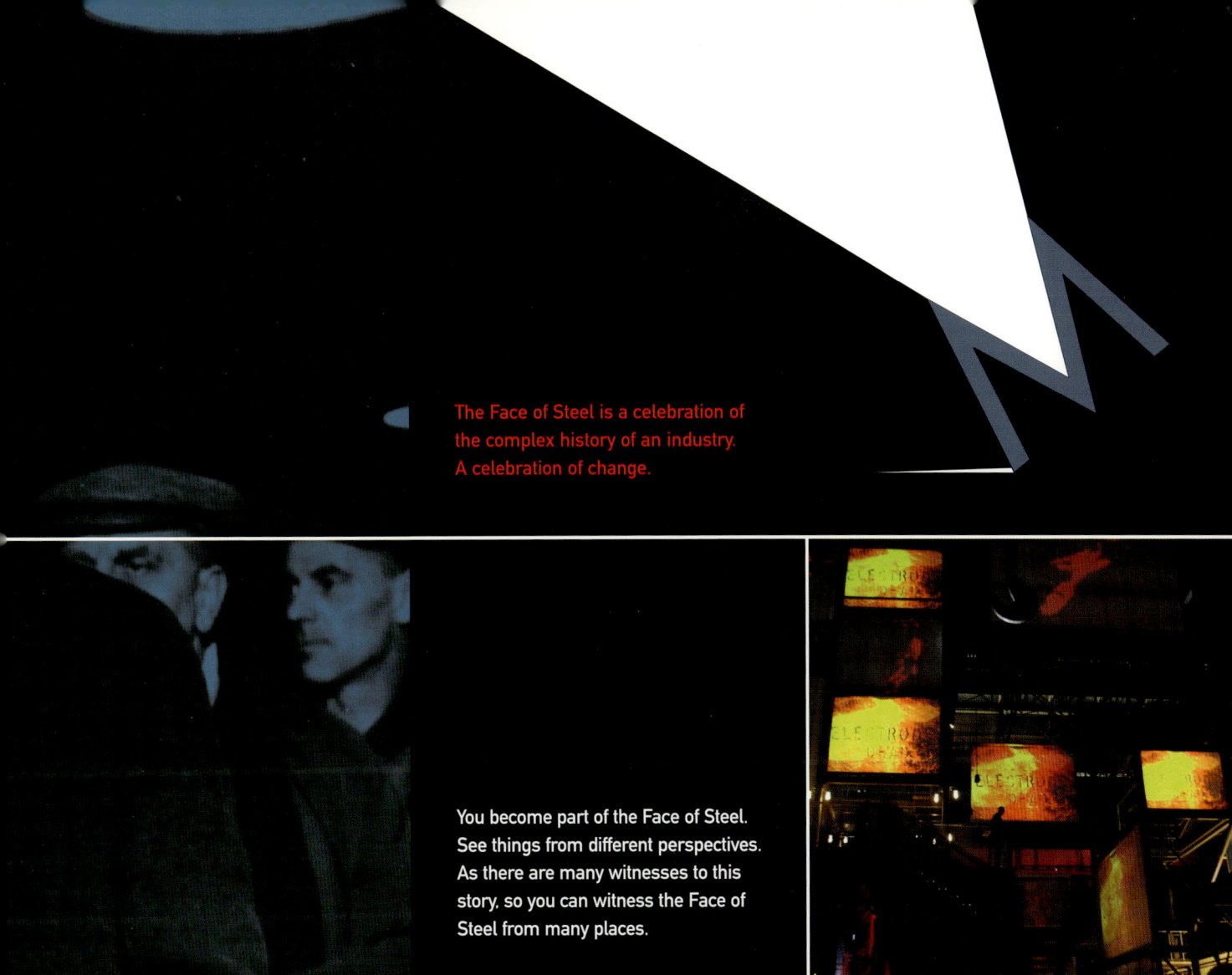

The Face of Steel is a celebration of the complex history of an industry. A celebration of change.

You become part of the Face of Steel. See things from different perspectives. As there are many witnesses to this story, so you can witness the Face of Steel from many places.

### THE PLACE
"Steel, Peech & Tozer's was a family affair... everybody knew everybody..."
Steel, Peech & Tozer – the company that built Templeborough – was a big employer and thousands depended on it for their livelihood.
"We always used to say... if you behaved yourself, you'd got a job for life."

### HEAT AND DANGER
Steelo's or Steel Peech's as it was known, had a good record for health and safety. But steelmaking was hot, hard, dangerous work.
"...clogs and goggles... sweat towels in their mouths... sweat just pouring down 'em."
"...one of the ladles exploded... it must have killed him – the shock of it... he was sat in the middle of the melting steel slag... burning..."

### GOOD TIMES
This was a deeply-rooted community, and both at work and at home, people looked after each other. Sharing the bad times – and the good…
"…footballing… swimming… dancing… Turkish baths on Saturday nights…"
"Blackpool. That was a works' outing, that was. 6 trains… going up at a time…"

### WAR WORK
There were always hardships. Always changes. And none worse or more shocking than the War and the Blitz. The men in the forces. Women doing work that was traditionally male.
"When the sirens would go, they'd rush to close the doors of the furnaces. So they wouldn't be seen from above…"
"We had lady crane drivers… good drivers they were… them women worked hard…"

### PRIDE
Despite the hardships, people enjoyed their work and took pride in it. Friendships and loyalties were strong.
"What I love about it are the people I work with…"
"…the lads have got tremendous pride in the work they do… we're a team."

### ACHIEVEMENT
Templeborough was a hugely successful steelworks. At its peak, it was more than a mile long and employed as many as 10,000 people.
"There were 6 arc furnaces all in a line. The target… was to make 1.8 million tonnes a year of steel… enormous… just enormous."

### TRIUMPH AND TROUBLES
The industry was nationalised. Steelworkers broke 60 production records. But competition was growing. Too much steel was being made. Job cuts led to a strike.
"We broke a production record… a monthly, a weekly and a 24 hour record…"
"We were on strike for 3 months… we came out of it with nothing."

### MOVING ON
For Templeborough, high costs, over-capacity, foreign subsidies and an unsympathetic government all proved too much. Though productivity had risen consistently for 14 years, in 1993, this great steelworks was forced to close.
"From those who fought so hard to make the plant a success, there's disbelief and despair."
"Templeborough has been an absolutely unbelievably good plant… one of the best in Europe… there aren't any better."
From 1979-1993, output per man in the UK steel industry rose by more than 300%. Steel continues to be made in the area, with great efficiency, and other, new industries are bringing new life to the community. Starting new chapters. Moving on…

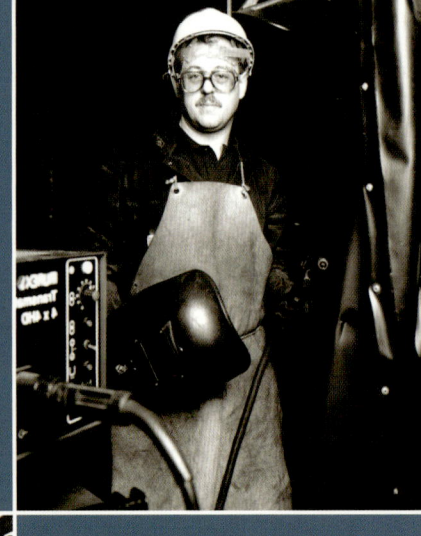

Bus conductors. Burns and worse.
Grandad's views and bathtime.
Mums at work. Choose your dirt.
Snow and tatties. Margaret Thatcher.
Somewhere totally different.

### THE VOICES OF STEEL
The voices you hear as you move through The Face of Steel are taken from hundreds of hours of recordings with people who, for the most part, either worked at Templeborough or were close to someone who did. What they have to tell us is fascinating, frequently very funny, often moving, always interesting. These are just a few examples from their amazing 'collective memory'.

"...a blast furnace at the top of the street... a coke oven at the bottom... and whichever way the wind blew was the dirt for the day."

"...grandfather believed in the women being at home... being kept short of money... and having lots of kids... 13 they had... can you imagine it? In a 2 up 2 down terraced house."

"We went on strike on the 2nd of January 1980... Margaret Thatcher was in power then... wanted to get at the miners. We were a test case and she smashed us easy. No problem."

"We didn't get a rise in money... two years later they started to shut the place down."

"It's sink or swim. It's about surviving."

## "You could... feel your chest pounding with the noise and vibration."

"In those days... Templeborough was the Mecca of steelmaking."

"At worktime, cheeky sod bus conductor used to shout – 'any more for the pleasure garden ?!'"

"They called it the Devil's Kitchen."

"Worst thing about working here was the heat."

"If we had a blow back... the men would get their hands and faces burnt... they'd look horrific."

"There are accounts of a person here stepping off the platform into a ladle of metal. Stepped off and that's the last you see of 'em."

"...snow never settled round Templeborough because of the heat from the furnaces."

"Father was first to be bathed... then mother... then the youngest... and me as the oldest child last... all in the same tub of water."

"I've only ever worked in melting shops... I don't know what else I'd do."

"I thought I knew everything about steelmaking until I went there. Templeborough was totally different."

"They closed us at the end of the week."

Magna would like to take this opportunity to offer thanks to everyone who contributed to our oral history.

# Fly through air

## Make clouds of smoke. Ride out a gale. Get close to a tornado. Be a great musician. Make crazy music. Be rude. Belch. Make wind and whistle.

**GET IN A SPIN**
This is you – in a freely rotating chair – behaving like a tornado to demonstrate 'centripetal' forces. Get spinning. Tuck your legs in, make yourself smaller and narrower, and you go faster. Stick your legs out, make more of yourself and you slow down. Simple, really. But revolutionary.

Compare this with other spinning exhibits. Dyson vacuum cleaners apply the same principles embodied in cyclones and tornadoes to help suck the dirt and dust out of your carpets.

You don't fly off a roller coaster looping-the-loop, because the cars are going fast enough to make the centripetal force holding them to the rails greater than the force of gravity which is trying to throw them off.

We can't see it, smell it or taste it. We hardly ever notice it. But there's 5,200,000,000,000,000 tonnes of it surrounding the Earth and we can't live or even breathe without it. This is the story of air and what it does. As you climb up onto the high steel walkway, it's the first thing that grabs your attention – suspended like a great translucent airship, frozen in flight, far above the ground. This is the Air Pavilion, where you can have a great time flying through air.

### CHASED BY THE WIND
From flat calm to gale force. Enter the Air Pavilion and you never know what you'll find. A rush of air surrounds you. Or is it following you? These are 'The Winds of Change' – one minute here, one minute there. Chasing you. One minute calm, the next a breeze, the next near gale force. This is sci-fi. Stay long enough and it'll catch you.

### BLOW SMOKE RINGS
Smoking gently. Greeting you as you move onto Air Level from the Transformer House is an exhibit that starts the story of air flowing. Push down on the rim of the big cylinder, compress the diaphragm and you create perfect mist rings that float away high into the roofspace. It's called 'Cloud Rings' – by the California-based artist Ned Kahn.

Winds turn and deflect to the right in the Northern Hemisphere; to the left in the Southern Hemisphere. This is called the Coriolis Force and comes about because the earth rotates east on its axis. Tornadoes are too small to be affected by it. So they blow any way they like!

### AIR TORNADO
Tornadoes are the most violent winds on earth – formed when updrafts of warm air push upwards through layers of colder air and start to rotate. But far from being a violent twister, this one is gentle – 5 metres tall and formed by mist – drawn upwards and set spinning by fans. To see a really scary tornado, visit the Fire Pavilion.

Hurricanes. Pacific typhoons. Indian cyclones and Australian willy-willies are all names for spinning tropical storms that form over oceans. The good news? They're slower than tornadoes. The bad news? They're wider and go on a lot longer!

### AIR FORCE
A first-hand look at what air can achieve.

Air can create the lift that makes heavier-than-air machines fly. In the form of hurricanes and tornadoes, it can wreak havoc. Knock down buildings and bridges. In the Air Pavilion, you can share the air, and find out what it's made of – and what it's capable of – for yourself.

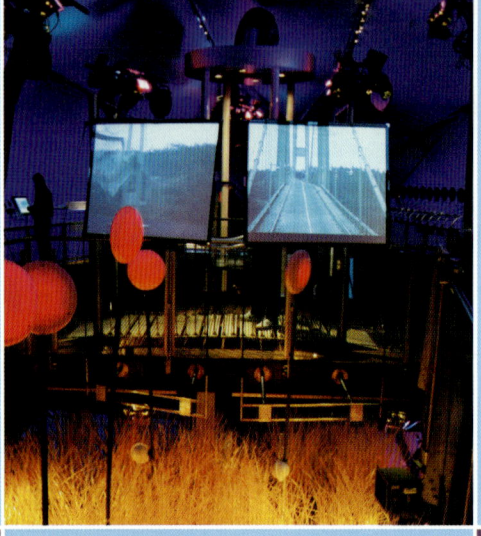

See early and often hilarious attempts at flight, in old photographs and film. Witness the role of air in steelmaking, and the history of air pollution around Templeborough. Discover how the famous Dyson Cyclone is no more than a storm in a vacuum cleaner. Uncover why Gertie galloped. Change the shape of 'Infalling Clouds' and create pictures in the mist. Play with 'Turbulent Orbs' and experience what it might feel like to control the Earth's wind and weather. Megalomania!

Air is 76% nitrogen, 21% oxygen, 1% water vapour, 1% argon, 0.03% carbon dioxide. The remaining 0.97% is made up of tiny amounts of neon, helium, krypton – look out Superman! – xenon and hydrogen.

### SWINGING BRIDGES
Steel is super-strong under tension. Ideal for suspension bridges. But things can go wrong. In 1940, the Tacoma Narrows suspension bridge over Puget Sound in the USA famously rose up and collapsed, when its road deck oscillated in the wind. Here you can investigate how bridges hold up in a wind tunnel. Stand on a suspension bridge while you watch the Tacoma Bridge fall down on eye-witness film. It's all right, though. The problem's been solved.

The Tacoma Narrows Bridge moved alarmingly from when it was first built. So much so that people called it: 'Galloping Gertie'. All suspension bridges move in the wind. Nowadays engineers make sure that the movement doesn't get out of control - or disaster strikes.

### PLAY INSTANT MUSIC
Ride the air waves. You can play stunning-looking steel instruments – like pan pipes, an air-powered flute, organ and steel drums – even if you have absolutely no experience of playing musical instruments. The sounds have been specially created so they sound good together. So you sound good, too. Or at least – we hope so!

Pan pipes were originally played by Stone Age people, with hollow canes of different lengths tied together. The longer the pipe, the longer the air takes to travel through it, the lower the frequency of the note. The shorter, the higher.

### WAVING FIELDS

The elegant, enigmatic Air Pavilion centrepiece. It looks like a field of corn. But what does it do? Blow hard into one of the fans. An electronic signal activates a larger fan below, which then blows complex, changing wave patterns into the corn. No aliens are involved.

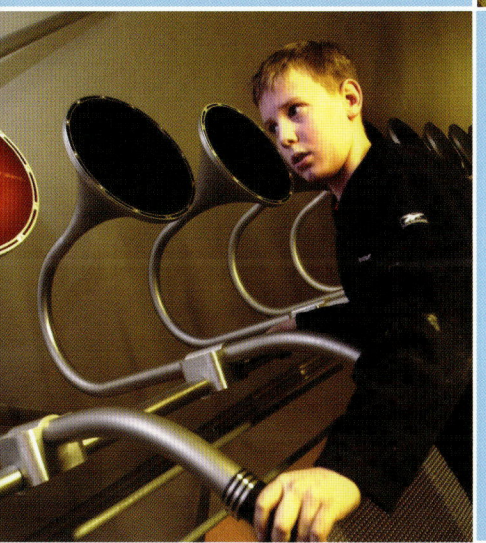

### TEST YOUR WINGS

Two transparent wind tunnels let you control the airflow around a wing. The air – in this case, shown as mist – flows faster over the wing than under it. This causes a difference in pressure – it's lower over than under – and this makes the wing 'lift'. Lift helps create flight.

### MAKE EXTREMELY RUDE NOISES

Go on. You're allowed. Our bodies need air to live. They use oxygen from air to 'burn' food for energy. They create waste gases. Make funny noises. Don't worry – you're not alone. We all do it. Here, 11 horns light up and play when you press their air bulbs – belching, sniffing, farting, breathing, whistling, sneezing, coughing, snoring, tummy rumbling, singing and speaking. Air in our bodies? Rude? You bet!

In a lifetime, you breathe a total of around 20 million litres of air. Sneezes travel at 150 kms an hour. At 93.7 decibels, a cough can sound as loud as a burglar alarm. The average person farts 15 times a day.

### FLY LIKE A BIRD

Another wind tunnel. This one lets you experiment with the 4 forces of flight – lift, weight, drag and thrust. Put your hands into the wings and feel the forces of the wind. Tilt them until you get the right balance of lift and drag – 'the angle of attack'. Then the floor seems to open up – and you're flying!

Modern aviation was born on 17th December 1903 at Kitty Hawk, North Carolina, when Orville Wright flew the biplane 'Flyer' for all of 37 metres. He was in the air for 12 seconds. Said his brother, Wilbur: "A bird's skill as a flier is not apparent. We only learn to appreciate it when we try to imitate it."

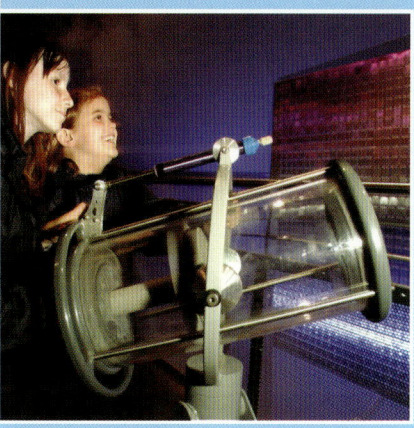

### AIR CANNON

Here, hundreds of small steel squares hang freely on metal hooks on a big, wall-mounted grid. You pump the piston to fire a bolt of compressed air. The charge hits the squares and creates a glittering wall of constantly changing patterns.

# Catch fire

**How would you like to stand next to a pillar of flame?** Play games with fire? Say hello to a few thousand volts of electricity? Melt steel? Race with an electromagnetic crane? Flirt with danger? Enjoy the fun of the forbidden? At Magna, you can do all of these things and more.

### A TORNADO OF FIRE
This is scary. You stand right next to it. Kerosene is injected onto the big metal base. It ignites. The flame roars upwards. Drawn and driven by fans, it twists and writhes – 5 metres high. This is how a vortex works. You can feel the heat. Hear the noise. Sense the power.

There are only 6 sources of heat – the sun, the Earth, fire, electricity, nuclear energy and friction.

This is about the awesome, primeval power of fire. It's also about how we control it and use it to power our lives. It's a frightening, fascinating, funny, high voltage experience.

### JACOB'S LADDERS

Start in the Transformer House and the air crackles around you. Sparks and flashes race across the ceiling. White arcs of light dance and flicker – these are Jacob's Ladders. Flasks filled with helium and argon gas blaze pink and blue, buzzing and hissing. Flames reach for the roof. This is 15,000 volts at work. Outside – above the walkway – a dancing trail of sparks leads you to the big, black box that looms in the shadows – the Fire Pavilion, where you can catch and play with the power of fire.

The electric artworks – Jacob's Ladders and Lightning Zappers – were created by San Francisco-based 'artist in electricity' Cork Marcheschi – to convey the fiery sound and light effects of high voltage.

### MELTING STEEL

You do this – replicating the way electricity is used to melt steel in arc furnaces – just as it was here at Templeborough. Wait for the countdown. Press the button. A high voltage spark arcs between an electrode and a steel plate. The heat energy melts the steel – steel melts at 1538°C! That's nearly twice the temperature of a wood fire – 800°C – and more than 8 times the heat needed to fry chips – 180°C. That's hot!

See the arc furnace, life-size and loud at the spectacular Big Melt Show.

### LIGHTNING, WILDFIRE, FIREBALLS AND DESTRUCTION

Bolts of lightning crash and crackle around the box. You can see and hear them 200 metres away. Inside, on screens, there's more lightning. Fire burns through 180°. It spreads like wildfire. Bursts into a fireball. Then dies away. There are charred remains and embers. Rust turns to bright steel, as we begin to harness fire. Using it to turn the chemical energy of fuel into heat energy. The power for every stage of our development from the cave dweller's cooking fire to steel making, space travel, the electronic revolution and beyond. Travel through the flames and enjoy finding out more about fire's fiery magic.

Humans are unique in being able to create and control fire. It rarely occurs naturally – unless started by lightning or volcanoes.

Go power mad. Control heat and cold. Make cables glow red hot. Paint heat patterns on the wall. Fuel machines with your hands. Change shape. Play fiery computer games. Walk through steel. See the future.

### FIRE ENGINES
Making heat work.
Hands-on experiences demonstrate how the heat created by fire can be made to work to our advantage. To give us power. Warm us. Cool us. Entertain us. Find out how fire performs. How heat behaves. Have fun with coils and cables. De-magnetise steel. Crank up a refrigerator compressor and watch it go hot and cold. Use body heat to change the colour of heat sensitive materials. Make engines go faster just by touching them. Solve the mystery of lava lamps.

### FIRE SAFETY
After creating all this fire, Magna brings you cures for fire too. An interactive computer game explores the home environment, identifying the best ways of ensuring fire safety. Proof that you can have fun playing safe!

### SEE CONVECTION
Adjust the heat of a coil. The coil heats the air, which expands and becomes less dense. It floats through the colder, denser air above – which then flows downwards to replace the warmer air. This is a convection current – now projected in flowing shapes on the wall. Strange and eerie. Check out other exhibits demonstrating expansion, conduction and radiant heat.

### STEELSAVER
This is for real. But it's also a game. A race to the finish. Steel is the most recycled material on earth. In a previous life, your baked bean can might have been something completely different. Steel for recycling is pulled from heaps of scrap by cranes equipped with powerful electromagnets. Now you and a rival operator can drive the cranes in competition. Switch the current on to grab the cans. Switch it off again to drop them in the hopper. The first to fill a hopper wins.

In an arc furnace, a 100 tonne load of scrap steel can be melted in an hour. Every year, 7 million tonnes of scrap steel – from things like cars, cookers, fridges and cans – are recycled.

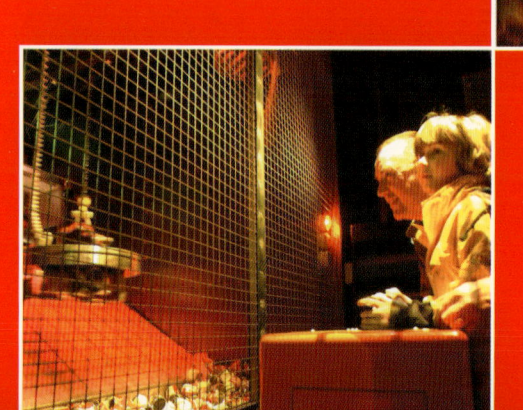

### THE MYSTERY OF THE LAMP
Lava lamps work by heating wax in salty water. The electric current heats the lamp. The lamp heats the wax, which rises, cools and drops, and is heated, rises, cools and drops... no cool home is complete without one.

### SHAPE SHIFTING

Casting, rolling, forging and machining – when it comes out of the furnace, steel is formed in many different ways. You can simulate casting with coloured fluids. Roll foam to see how steel is shaped into products like steel bars and railway lines. All sorts. Magna's forge demonstrates the forging of steel, with blacksmiths and artists staging occasional special events. Films show all these shaping processes from different times in steelmaking's history.

Steel is not only used to make cars, electrical products and cans. It holds up the World Trade Centre, flies on the Space Shuttle, drills into the heat of the Earth's core.

### FIRE WORKS

You add metal salts to a Bunsen burner – different substances create different colours. Sodium burns orange. Copper burns blue. This is the basis for fireworks. The change is irreversible.

When the Chinese invented fireworks in the 8th century AD, the first fireworks they invented were bangers. They were designed to scare away evil spirits!

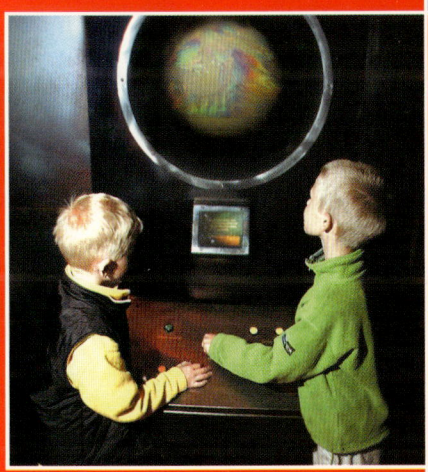

### COOL CRYSTALS

Control the heat supplied to crystals magnified on a projection microscope. The crystals melt and freeze as you add or remove the heat, just like ice crystals do. The change is reversible. The images spectacular.

The beautiful shapes of crystals have led some people to believe that they have healing powers – passing their perfection to the human body.

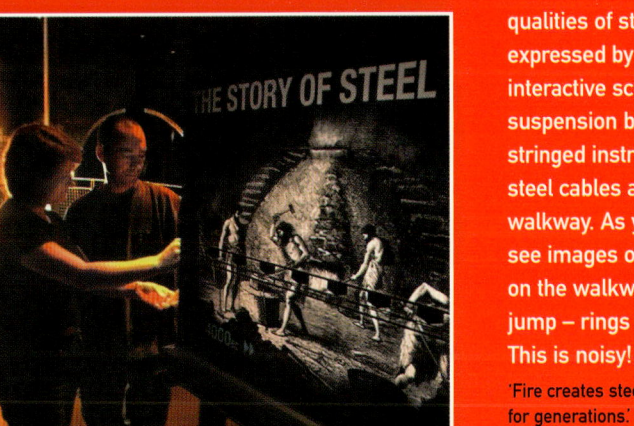

### FIRE AND THE STORY OF STEEL

No fire, no steel. No steel, no modern world. No cars. No trains. No planes. No kitchen appliances. No computers. Hi-fi's. TV's. You name it. This is 'The Story of Steel', told in 7 different time frames. Just by moving a computer plasma screen along a timeline, you can reveal film, video, photographs, illustration and graphics that picture the past and the present of steel.

### THE AMAZING PROPERTIES OF STEEL

Steel comes out of the fire molten, blazing and amazing. Strength, workability, wearability, stainlessness, resilience, versatility – all the amazing qualities of steel are powerfully expressed by artist Gerry Judah in an interactive sculpture that looks like a suspension bridge or some weird stringed instrument, combining film, steel cables and a sprung steel walkway. As you walk through, you see images of the future. Your weight on the walkway – especially if you jump – rings rows of steel bells. This is noisy!

'Fire creates steel. Steel can survive for generations.'

# Dive into Water

**If you want to make a splash.** Get soaked. Track a salmon. Tell truth from falsehood. Make liquids dance. Explode and fire a rocket. Chill. Bubble. Or pee. Then go right ahead.

### THE BIG BLUE
Hundreds of laser cut, stainless steel ripple and wave forms flow across the ceiling and cascade down the walls, guiding you from the Transformer House to the Pavilion itself. Lighting makes the steel appear to move like eddies in a river, the swell of the sea. 'The Big Blue' by Sheffield-based artist, David Mayne.

### WELCOME TO WATER WORLD
This is a world where water runs, drips and gushes from the walls – from the landscape – into big stainless steel tanks. The tanks are constantly overflowing. Water is pouring beneath your feet. It's all around you. Images of the sea, storm clouds, rain and rivers are projected on steel, reflecting and distorting the pictures into watery abstracts. Then – and this is well worth waiting for – real mist rises from the water and the rippling surface is broken by rain – real rain that can splash your face. The rain passes and the cycle begins again.

Only 2% of the earth's water is fresh. Four fifths of the fresh water is trapped in glaciers and the polar ice caps. One third of the rest is contained in Lake Baikal in Russia. All life on land depends on 0.4% of Earth's water.

We drink it. Wash in it. Swim in it. Depend on it. It's 60% of our bodyweight. 70% of the Earth's surface. It carved the Grand Canyon. Creates tidal waves, floods and monsoons. Now you can get inside it. Throw it about a bit. Get wet. Look down from the walkway and you'll see it – a building shaped like a wave, outlined in electric blue. It's the Water Pavilion where you can dive in, have fun and make some incredible discoveries.

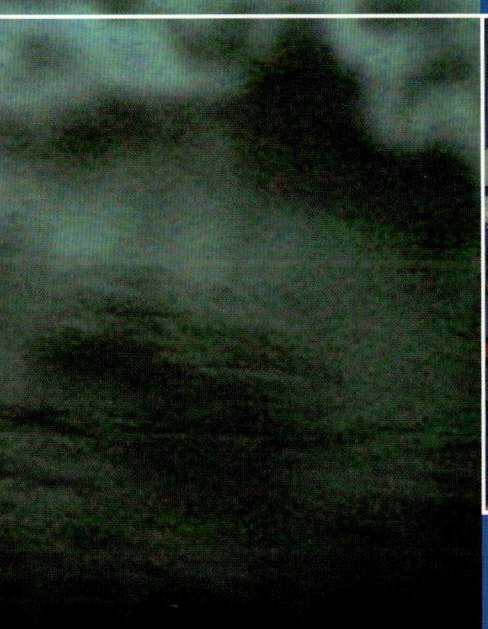

### SUPERSOAKERS – SPLAT HUMAN TARGETS!
Water has enormous energy potential. As you're about to discover. Get hold of one of the super-size Supersoakers. Aim at the targets or real live human visitors passing behind transparent plastic-covered portholes. Pull the trigger. Release the pressure – the energy in the water – and splat! You've got 'em. Now do it again.

### FIRE A HYDROGEN ROCKET
Cranking up a generator creates electricity, which turns water ($H_2O$) into an explosive mix of hydrogen and oxygen gas. Press 'FIRE'. A spark ignites the gas and the reaction shoots the rocket high into the air. Water exploding!? Who'd have thought it!

### COOL STEEL
Squirt water onto a very hot steel plate. What happens? It dances, hops and skates around. What's going on? The cool water boils and quickly evaporates. The evaporation creates a cushion of vapour and the water floats and bounces about on it. As more water evaporates the process speeds up and gets turbulent. The turbulence sends the water flying around.

When water comes in contact with molten lava from volcanoes it evaporates instantly. The effect is explosive – like a water and lava grenade.

### PARTING WATER
A steady jet from a pipe creates a seamless dome of water. Mist is trapped inside. The touch of a hand – something placed over the pipe – releases it in billows. Objects placed in the flow create strange shapes. You're sculpting. Carving water.

Squirt. Spurt. Follow cycles. Turn wheels. Sharpen knives. Bang hammers. Find out how much water it takes to make jeans. Or create a landscape. Lose yourself in a vortex – and don't forget to flush. Be naughty with water.

## BLUE WHIRLPOOL

Beautiful and fascinating, the whirlpool works in two cycles. The first creates a smooth flowing vortex. The second is unstable, oscillating violently, with one vortex breaking away and another forming – and another. All lit by blue neon.

The slightest movement, surface irregularity or the angle of a tap or pipe can change the direction in which water swirls down a plughole. Contrary to popular belief, whether it's clockwise or anti-clockwise has nothing to do with whether you're in the Northern or Southern Hemisphere.

## WATER WORKS

All sorts of fun ways to mess with water, find out how it works – and make everyone very wet! Water is the only natural substance that occurs as a liquid, a solid and a gas at normal temperatures. Experience them all. Walk through water and mist. Thaw ice with your hands. Measure your weight in water. Work out how much you and other people need. Use water and steam power. Make waves. Power a lighthouse. Float a barge up and down canal locks. Put your head in a monster's mouth. Sculpt water and be amazed by great big bubbling geysers. We provide waterproofs and towels. But no-one gets out of this dry!

What's a geyser? Heat from volcanic rocks causes underground water to boil. The boiling water pushes to the surface through cracks and blasts its way out as steam.

## WATER WHEELS

A big stainless steel water wheel, part fed by overhead jets, uses wheels and pulleys to power a grindstone and a tilt hammer – once used around Sheffield for sharpening knives and forging steel. You can practice your sharpening technique, power the hammer, play games with 4 kinds of water lifting device and test out 3 other water wheels – each of a different type – to see which is most powerful.

'Archimedes Screw' is a water lifting device which may have been invented by the eponymous ancient Greek scientist, Archimedes, in the 3rd Century BC. It's still used today by farmers in developing countries to lift water from irrigation channels onto their crops.

## FOUNTAINS OF KNOWLEDGE

We live in a world where manufacturing industry consumes vast amounts of water to make everything from aeroplanes to potato crisps. The Fountains give you the opportunity to estimate, establish and compare exactly how much is used for each of a huge variety of products. Just turn on the taps and keep your eye on the float.

It takes 750,000 litres of water to run a power station for a day; 150,000 to make a car; 90,000 to make a tonne of steel; 15 to refine a litre of petrol; 1 to make a small bar of chocolate. How much for thousands of other everyday products?

#### WATER CYCLING

A working model of water's lifecycle, where it evaporates from the sea, merges with cold air in inflatable clouds, turns back to water, rains on the hills, pours into river and streams, flows down to the estuary and into the sea, where it evaporates...

In 1985, in the Yorkshire town of Thirsk, it rained winkles and starfish – probably sucked up by a water spout in a thunderstorm and dumped there – 72 kms from the sea.

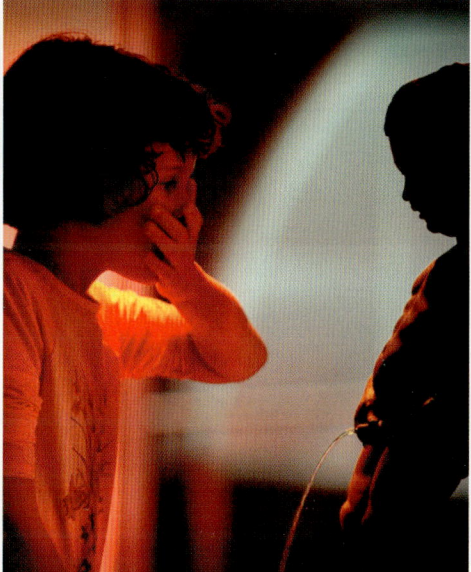

#### WET PLAY

Interconnected pools and waterways designed for under 7's of all ages to play with. There's a pond where you can float things and sink things, squirt and blow bubbles. Flowing water, where you can build a dam with sandbags, operate a sluice gate and meet a sea monster. Plus a flume, rope and bucket pumps, a spout, a wheel and whirlpool. All powered by water from a giant tap floating in the air.

#### TELL FISHY TALES

Fish – and salmon in particular – are good indicators of water pollution. With the help of a computer and a big screen, steer a salmon up the River Don through 4 different periods of history, where hazards may vary from herons and fishermen to sewage and toxic chemicals. Nowadays, though, things are definitely getting better – and you can see all this from the fish's point of view. At least you can – if you survive long enough!

The Don was once said to be 'the dirtiest river in Europe'. Now it's being cared for. Fish are being introduced and by the year 2050, salmon may spawn there again.

#### MIND YOUR P's AND Q's

How much do you know about the water in your body? How it gets there. How you use it. How you get rid of it. Answer true or false to the questions on the screen and Brussels' most famous statue – the Mannequin Pis – reveals the answer by peeing – with astonishing accuracy – into true or false buckets!

Your blood is largely water. It carries waste from your body to your kidneys, where 1,500 litres of blood are recycled every day. Tiny filters called 'nephrons' collect the waste which combines with water to form urine – which is passed in the usual manner.

#### FLUSHED – DOWN THE PAN!

A toilet, no less, with a display on the door that reveals how much water has been used at Magna since it opened and how much has been used today. Find out how many toilet flushes that adds up to. But whatever you do, don't open the door – there could be somebody in there! A third of all the water you use is flushed down the toilet. Each flush equals 7.5 litres. We use 70% more water in the home now than we did 30 years ago – but the amount of water available for use remains the same. 'All life depends on water.'

# Move earth

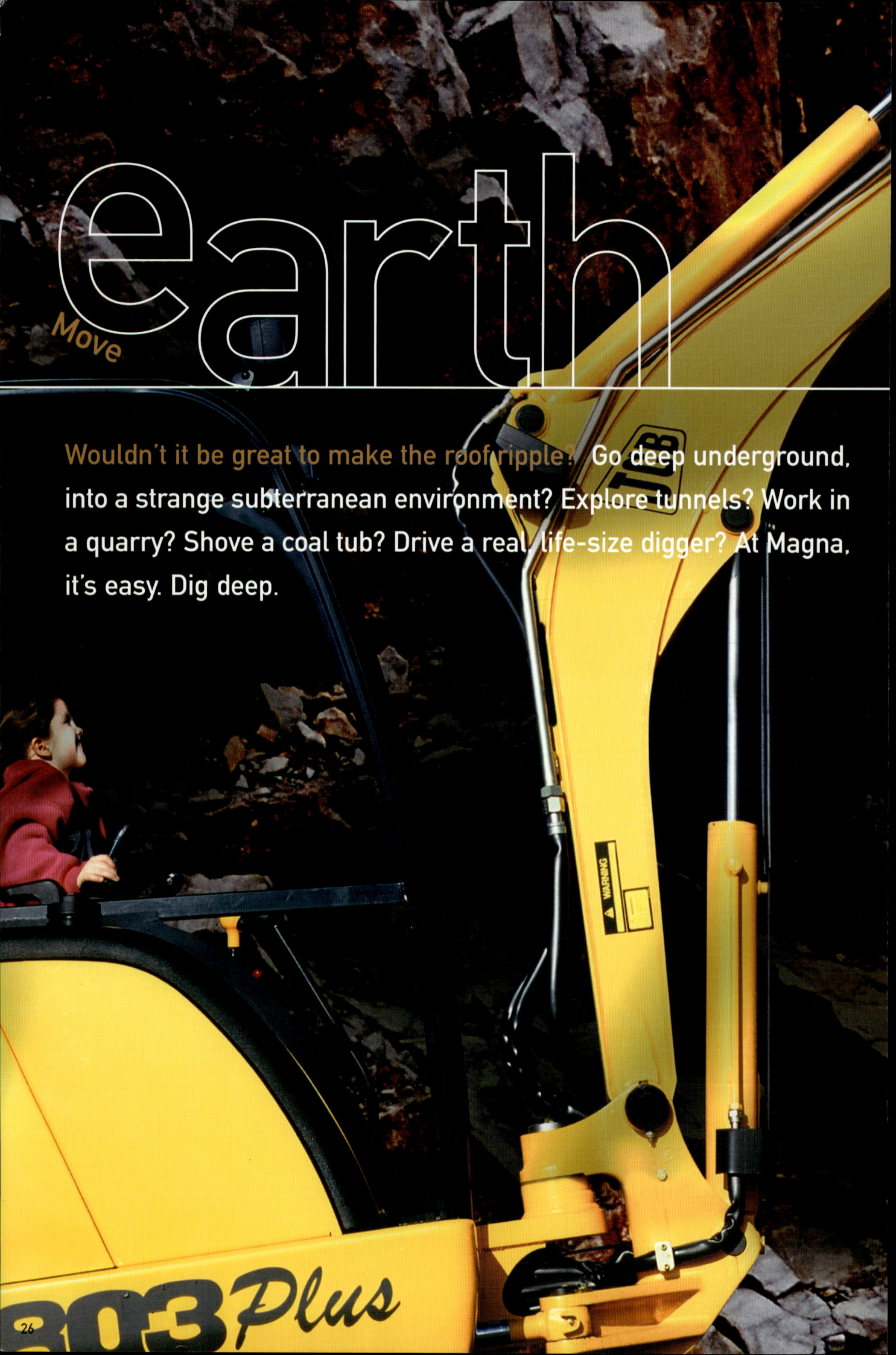

Wouldn't it be great to make the roof ripple? Go deep underground, into a strange subterranean environment? Explore tunnels? Work in a quarry? Shove a coal tub? Drive a real, life-size digger? At Magna, it's easy. Dig deep.

Get in and amongst it. Push it around. You're buried in it. Find out what we get out of it. What it's made of. How we work it. Need it. Need to care for it. How great it is to play with. Down and dirty.

### EARTHNET
You're at Earth Level in the Transformer House. About to be stalked. The Earthnet's sensors spot you and start to follow you around. The kinetic, elastic spider's web ceiling is in constant motion – sensing you, looking for you – undulating, tracking your movements like a spider seeking its prey. The surface is moving up and down like some weird living creature – testing the air. Are you there?! Earthnet is by Athens-based artist, Toby Short.

### RIDE THE UNDERGROUND
You're going deeply subterranean now – under the earth into the deep, dark depths of the great steel works. Under huge girders that once supported colossal machinery roaring and belching fire above – the melting pot of 'F' furnace – one of the six that were here. Can you imagine the heat? Feel the weight of the steel and the earth? Breathe deeply. And carry on.

### DRIVE YOUR OWN DIGGER – A REAL ONE – NO LICENCE NEEDED
In the Magna quarry, there are 3 real life-size JCB diggers, complete with fully equipped cabs and all the controls. Hop in. Operate the digger arm. Fill the hopper with rocks. Dig alone – if you want to – or compete against friends and family. The winner is the driver who moves the most rocks against the clock. A scoreboard records the top score each day. These are not toys. They're the genuine article. Dig it.

JCB was founded by Joseph Cyril Bamford in 1945 – creating the first 'backhoe loader' in 1953. The company is now among the world's top five construction equipment manufacturers; JCB is in the dictionary, and there's even a 'Dancing Digger' display team. Cool – or what?

### TUNNEL IN
You enter the Earth Pavilion by tunnels. One you walk through. One children can crawl through. One you can push a coal tub through. There are peep holes where you can see animal life underground. Yell down a mineshaft: 'Is anybody down there ?!' See the earth forming and moving in abstract, projected shapes – fire and rock at its core. You're surrounded by the sights, sounds and smells of earth – here to dig and delve into its secrets. Play with the dirt and rocks.

Try pushing a coal tub the same size and weight as those pushed by children in the mines around Rotherham in the early 19th century. Called 'hurriers', these children rarely saw the light of day 'except on Sundays'. Boys and girls under 10 – and women – were finally banned from working underground in 1842. The tub doesn't feel that heavy. But try pushing it all day. Every day.

27

*Smash rocks. Excavate. Pick stone. Pull levers. Hoist your friends and family. Go down the mine. Have a rest. Decide the fate of an industry. Save the world.*

### SEE GIANTS AT WORK
Some of the world's largest and most powerful machines are needed to quarry rock. See them at work on video, at Shapfell in Cumbria – quarrying limestone for steelmaking – and at Tunstead in Derbyshire – producing soda-ash for glass. They're massive. Awesome. Humungous.

### SORT ROCKS
Sorting and grading machines are used in quarries to separate rocks into different sizes. Here you can help sort them as part of the continuous conveyor system.

### FANTASTIC INTERACTIVES
All the Magna Pavilions have fantastic computer interactives – and Earth is no exception. As well as the story of coal, you can discover how and when the Earth's mineral resources were formed. How the rocks needed for steel making were laid down millions of years ago. How the industry was based on local materials, but now brings them in from all over the world. And how the Earth's resources are being used up at a rate we can never sustain. What does the future hold? Can technology save us? Who knows? It's up to us to decide. Or will we have to live very different lives?

Mining the Earth's minerals provides most of the raw materials of our daily lives, from steel, to plastics, glass and buildings. Each of us uses around 18,000 kgs of new mineral materials every year – well over 1,250,000 kgs in an average lifetime.

### LEVERS, PULLEYS, GEARS AND HYDRAULICS. THE POWER TO MOVE YOU
These are the basic tools for helping us do work, lift loads and move things more easily. You can operate them all and see how they work, by moving rocks around on high level conveyors. Installations at a lower level let you test how much effort – or how many people it takes – to lift rocks with different length levers, different numbers of pulley and different types of gears. You can also use a hydraulic cylinder to see if you can raise a platform with up to 4 people sitting on it – you might be stronger than you think.

Archimedes (287-212 BC) said: 'Give me a lever long enough and a fulcrum on which to place it, and I shall move the world.'

### EXPLODE A ROCKFACE
You're in control. On video is the wall of the quarry. Drill-holes contain charges of liquid ammonium nitrate and fuel oil – they don't use dynamite much any more. The holes have been plugged with crushed stone – now it's down to you to detonate the explosion. You push the plunger. There's a terrific roar and all hell breaks loose as rocks and rubble fly into the air. Cascading rocks crash onto a plinth, ready for loading and sorting. How are your ears?

A large quarry may routinely blast 100,000 to 150,000 tonnes of rock per shot of explosives.

### ARE YOU A NIMBY?

You should be wearing a helmet. Now look – and listen. They're redeveloping a coal mine in your neighbourhood. Should they continue deep mining or go open cast? Open cast takes coal close to the surface. It can be a lot cheaper and more productive than deep mining, but scars the landscape and can cause other social and environmental damage. Video puts both sides of the case. Then you vote. You stand in a 'Yes' or a 'No' box and literally vote with your head – the camera recognises and registers the reflective strip on the top of your helmet. Will you vote 'Yes' to open cast or will it be a case of 'No' thank you – 'Not In My Back Yard'?!

The world's deepest mine is the Western Deep Levels gold mine in South Africa – at 4kms below ground. The enormous Carajas open mine in the Amazon jungle is the first iron ore mine to win an international award for good environmental management. It produces a quarter of all the world's iron ore.

### EARTH SHATTERING

How we work and enjoy; use and abuse the Earth's resources. This is a work experience – all fun and no wages – where you get to explore the way we handle some of the world's huge mineral wealth. You can work the machines. Feed the conveyors. Try levers, pulleys, gears and hydraulics. Play games. Go backwards and forwards in geological and industrial history. Take a break. Listen to short stories. Think about how we might look after the Earth better in future. Why not start by doing some work?

### USE THE BIG BUCKET WHEEL EXCAVATOR

Excavators are one of the key ways of moving large amounts of earth and rock. Try your hand at it with Magna's big excavator wheel. Scoop up the rocks, load them and send them for sorting. It's easy really.

Bucket wheel excavators are among the largest mobile machines on land. The biggest can move 12,500 cubic metres of rock an hour – that's more than 250 double decker buses full.

### BLACK GOLD

This area is famous for steel. It's also famous for coal. Now, through a large screen, interactive game you can experience the hard, dangerous work of a deep coal mine through 4 different periods of its history. See how it changed as machines came to take over the heaviest work from the miners, in much the same way as happened – in a different industry – to the people of steel.

### REST. SIT COMFORTABLY AND HEAR A STORY OR TWO

Or five or six. The Earth Pavilion has a Rest Hut for small children (or perhaps their parents!), who could do with a break from all the excitement. Push button controls call up a choice of stories. Then, maybe, you should retire to the O$_2$ restaurant for a glass of pop! 'Nearly everything we use comes from the earth.'

29

# The big melt

**Be prepared to be shocked. Amazed. Mind-boggled.** Avalanched with noise. Experience the electricity. The lightning. Smoke and flames. The roar and rumble. The awesome power of the mighty arc furnace.

**THE BIG MELT**
This is how the melting shop might have been when it was working flat-out. Less the heat. It's quite a show. As close as you're likely to get to the everyday experience of the Templeborough steelworkers. The noise and effects are tremendous. And remember – this is just a representation of one furnace. Originally there were six.

You're on the walkway again. Between the Transformer House and the Fire Pavilion. Looking down into the heart of the great steelworks. What you can see in the pit below are the remains of the gigantic 'E' Furnace, looking like the broken body of some enormous monster. You've just heard an announcement calling for your attention. The Big Melt is about to begin.

### 'E' FURNACE COMES ALIVE

This is a representation of the melting process in 'E' Furnace – created in sound, smoke and effects. That hum you can hear – rising, crackling, buzzing – represents 33,000 volts of electricity. The booming, crashing sound – like metal-on-metal – is designed to convey the furnace lid swinging clear and the colossal 'charge' of tonnes of scrap steel, limestone and pig iron thundering in. You can hear the lid swinging back. See what looks like three giant graphite electrodes moving over the metal, before being lowered through holes in the lid. Imagine the power peaking and a lightning-size bolt of electricity flashing from one electrode to another. Creating huge temperatures. Melting cold steel. Flames shoot upwards. Sparks fly. There's a bright blue light like an oxygen lance. Smoke everywhere. The walkway is vibrating. The din continues. Now there's a second charge. In reality, when enough metal was melted, the impurities were poured off and the steel flowed like molten lava into a giant casting pot on the other side of the furnace.

Catch your breath. Over the page, we'll explain what goes on during the Big Melt in a little more detail. But first, let's try and paint you a picture of the scale of the melting shop, what it looked like and what it could do with all six furnaces in place.

### TWO ENDS TO THE MIDDLE

The quarter of a mile long melting shop was part of a steelmaking complex five times longer. It was so big that people navigated by 'landmarks'. For instance, down in the direction of the Fire Pavilion, past 4 more furnaces, was the area known as the Sheffield End. In the other direction, beyond 'F' Furnace and the Transformer House, was the Rotherham End – where the molten steel was cooled and cast into giant 'billets', ready for finishing and shipping to manufacturers all over the world. The melting took place bang in the middle.

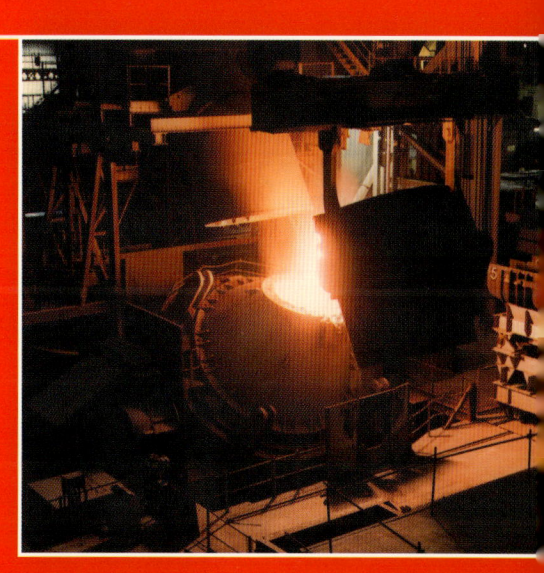

### ROARING AROUND THE WORLD

In its heyday, Templeborough was the world's largest electric melting shop, producing a quarter of Britain's electrically melted steel. Over 1¼ million tonnes of cast ingots a year. The works produced engineering steels – the flesh and bones of car engines, gearboxes, transmission systems and load carrying bearings. Railways across the globe ran on wheels and rails made from its steel. And Templeborough pioneered many techniques which set the standard for manufacturing engineering steels worldwide. So in a very real sense, wherever you went in the world, you could hear the echo of its roaring arc furnaces. For more about the Big Melt, see over.

The arc furnaces are installed. Templeborough leaves the rest of the world standing. A step-by-step beginners' guide to melting recycled steel. The thanks of a grateful nation.

## POWER ON
First comes the power – that hum you can hear – flashing from the control room in the Transformer House. At full production, Templeborough gobbled up 750 million units of electricity – roughly the same amount as a medium size town like Rochdale.

## COOKING UP A STORM
Among its many amazing qualities, steel is the world's most recycled material – and what you're seeing with the Big Melt is scrap steel being melted by the great arc furnace for recycling into all sorts of industrial and consumer products. It was a process that, at times, must have seemed like hell.

## A WORLD RECORD
In 1977, 'E' Furnace smashed the world record for the amount of electrically made steel produced in a week. Generating tonnes more liquid steel than any other furnace. Anywhere. For the men who worked it, coming to grips with 'E' was like standing on the rim of a volcano. The roar made your head spin and your ears ring. The heat knocked you backwards. Sucked your breath away. Sapped your energy. Get too close and its fury would eat you up.

## ARC FURNACE BEATS OPEN HEARTH
Steelmaking had always been a backbreaking, dangerous business, and for 70 years, the company that founded Templeborough, Steel, Peech & Tozer – Steelo's – expected its men to toil long and hard to stoke the hungry mouths of the old, smoke belching, coal-fired open hearth furnaces. Demand for cars and electrical appliances drove up demand for cheaper steel. Cheaper steel meant faster production. How was Steelo's to achieve this? In the early 1960s, the old furnaces were replaced with electric arcs – and at the flick of a switch, everything changed. Though jobs were lost, conditions improved. The output that 14 coal furnaces struggled to produce in 12 hours, 'E' and the five other new furnaces could reach with volcanic ease, in less than 90 minutes. How did they do it?

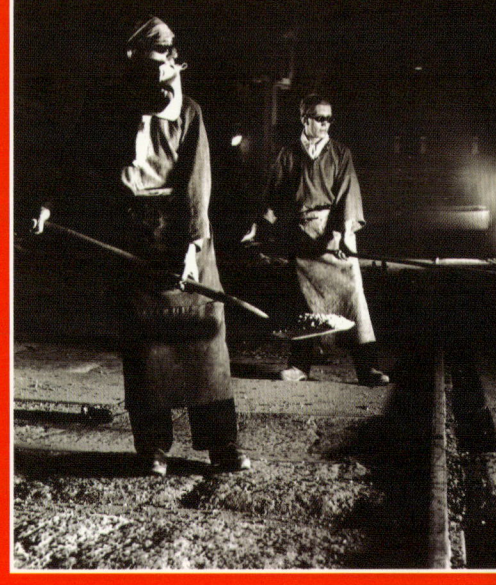

### TILT AND TAP

The furnace tilts. It's tilted one way to tip a thick skin of 'slag' congealing on the top into a 'slag pot'. Then the other way to 'tap' the liquid steel into a giant ladle, ready for further processing. That's 110 tonnes of pure steel.

### PIG AND MIX

Next comes the scrap – a mix of used steel that has to be carefully balanced and graded. Heavy stuff like girders. Down to industrial waste, metal shavings, nuts and bolts. It's mixed with limestone to soak up the impurities. And pig iron to aid combustion. The scrap in the scrap basket is pre-warmed by recycled heat from the furnace. The furnace is opened. The basket crane moves into place and in each of two charges, 80 tonnes of metal are dropped into the 'tundish'.

### THE ARC AND THE LANCE

The furnace is secured and the current let loose. Its coursing from electrode to electrode generates a temperature of 3000° Centigrade. Unimagineable heat. Alloys and refining agents are added through a door in the front of the furnace. The lance injects pure oxygen, speeding up the process. Raising the temperature. Removing unwanted gases. Steel samples are taken by a ladle on a very long pole – the temperature in there is now 1600°C

### THE PM's THANK YOU

The world record, set back in that December of 1977, was an average output of 74 tonnes of steel an hour. Every hour. For a week. The melting shop manager and his team got a Christmas card soon after. From the Prime Minister, Jim Callaghan. Thanking them for their Christmas present to the nation's economy. For making so much of the steel that carries our world on its back. Our buildings, our bridges, our industry, our technology. Thanks, then, to Templeborough.
By the time of the last melt in 1993, the one remaining furnace – 'E' Furnace – could produce 125 tonnes of steel in under two hours. Nowadays, arc furnaces can produce 100 tonnes an hour.

# history

**Swords. Nails. Tools. Ships and railways. Primitive. Old and new.** There's been metal working on this site for 2 millennia and possibly even longer.

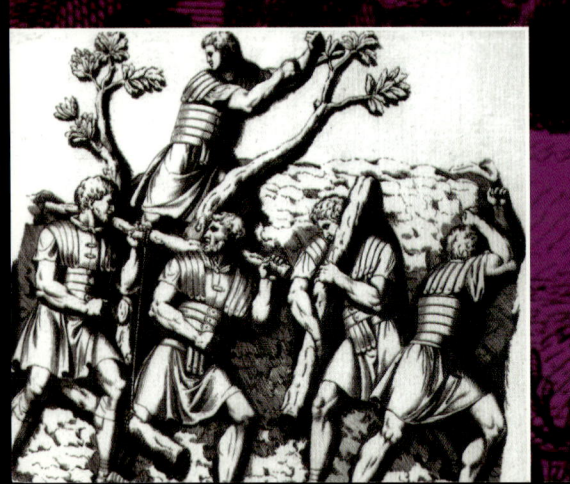

### 2,000 YEARS AND COUNTING
Out of 20 centuries of history, these are some of the key points and highlights for Templeborough, the steel industry, Steelo's and the melting shop.

### ROMANS, MONKS, GHOSTS AND GHOULIES
**AD 54 THE ROMANS ARRIVE**
They build a fort at Templeborough, with iron workings outside. People still claim to see ghosts of Roman soldiers.

**1161 BLASTED MONKS**
Just up the road, in Kimberworth, medieval monks make iron by smelting local iron ore in a primitive furnace.

### GUNS, BRICKS AND BLISTERS
**1608 FIRST BRICK BLAST FURNACE**
Opens in Rotherham to make nails. South Yorkshire has lots of iron ore. Plenty of trees to make charcoal for fuel. Rivers to provide water for cooling and power machines. More furnaces follow.

### 1709 STEEL FIRST MADE IN SHEFFIELD
They surround iron with charcoal in a conical furnace and bake it for 2 weeks. The resulting 'blister steel' is ideal for knife and tool blades.

### 1742 HUNTSMAN'S CRUCIBLE
Improves on blister steel by remelting the metal in coke-fired crucibles. Benjamin Huntsman discovers this in Attercliffe – 2 miles from Magna.

### 1748 ROTHERHAM'S FIRST STEELWORKS
Makes crucible steel – which turns out to be just the stuff for high quality products like guns.

## BESSEMER AND BRUNEL
### 1856 BESSEMER'S BRAINWAVE
British inventor, Henry Bessemer, invents the Bessemer Process, using a pear-shaped vessel for melting and compressed air to burn out the impurities. For the first time, large amounts of steel can be made cheaply.

### 1858 THE BIG BOAT
Rotherham supplies iron plates for Isambard Kingdom Brunel's 'Great Eastern' – the world's largest ship. Until the Bessemer Process takes over, steel is too expensive for big things like ships and railways.

## STEEL PEECH'S, WAR AND WOMEN
### 1883 STEEL, PEECH & TOZER
Company founded by Rotherham steelmen, including William Peech – father of Jim Peech, the man behind the building of Templeborough.

### 1892 OPEN HEARTH
Steelo's builds the first open hearth furnace in Rotherham. Open hearth takes over from Bessemer Process. Achieves high temperatures. Produces higher quality steel.

### 1913 STAINLESS
Sheffielder, Harry Brearley invents the most famous steel product of them all – stainless.

### 1914 WAR
The First World War explodes across Europe. With the men away fighting, women are employed for the first time to do some of the heavy work at Steelo's.

## TEMPLEBOROUGH AND THE 14 SISTERS
### 1916 TEMPLEBOROUGH BUILT
Artillery shells are in such short supply that they're rationed. Jim Peech convinces the government that Templeborough should be built to make steel for more.

### 1917 THE FIRST MELT
The first Templeborough furnaces are now melting steel for shells used in battles in France and Flanders.

### 1918 PEACE AND CHIMNEYS
By the time of the armistice in 1918, 11 furnaces are working. 3 more are added. Their 14 chimneys become known as the '14 Sisters' – famous Rotherham landmarks for the next 40 years.

Wars. Stoppages. Successes. Bombs. Nationalisation. Rationalisation. Foreign imports. Investment. The lack of it. World record. The last cast. Magna.

## UPS, DOWNS AND TURNAROUNDS
Steelmaking is a tough business. It's experienced tough times. Often, the victim of events beyond its control. Here's what happened at Templeborough.

## STEP AND GO
### 1919 LAND FIT FOR HEROES
With the First World War over, people look forward to 'a land fit for heroes'. It comes to nothing. The world economy slumps. European steel imports threaten Steel, Peech & Tozer's business.

### 1926 THE GENERAL STRIKE
A 9-day all-out stoppage brings the nation to a standstill – Templeborough included.

### 1929 WALL STREET CRASHES
Panic on the world's stock exchanges. But United Steel Companies – the group of companies to which Steelo's belongs – makes plans for the future. Templeborough has a key role to play.

### 1935 CONDITIONS IMPROVE
Steelo's invest in new plant and machinery. Improved working conditions. This year one of the UK's first contributory pension schemes is introduced.

## MORE WAR AND AFTER
### 1939 WORLD WAR TWO
In the 6 years of conflict to 1945, Templeborough produces 3 million tonnes of steel for shells, guns and the floating 'Mulberry' harbours used for the D-Day invasion. Women are doing the 'men's jobs' again!

### 1940 BLITZ
The Luftwaffe bomb Templeborough. Enemy propagandist Lord Haw Haw claims that the '14 Sisters' – the melting shop's famous furnace chimneys – have been destroyed. In fact, just the roof is damaged and only the threat of arrest stops Shift Superintendent Tubby England from restarting the furnaces before it's repaired!

### 1951 NATIONALISED
The post-war Labour government nationalises the steel industry. 10 months later, the new Conservative government denationalises it.

## GOING ELECTRIC
### 1959 PROJECT 'SPEAR'
The 1950s sees steel output increase steadily. Now comes a big leap forward. 'Steel Peech Electric Arc Reorganisation'. Steelo's open hearth furnaces are to be replaced with new electric arcs. At Templeborough 6 of them will finally flatten the '14 Sisters'.

### 1965 SWITCHED ON
The last open hearth furnace is 'tapped' for its load in December 1964. The last of the 6 new arcs is switched on in February 1965. Templeborough becomes the world's biggest arc plant.

## AGAINST THE ODDS
### 1967 NATIONALISED AGAIN
The Labour government nationalises the industry – yet again. Steelo's becomes part of British Steel Rotherham. What will happen now?

### 1973 RATIONALISED
Worldwide over-production, falling demand, foreign subsidised steel all combine to force capacity and job cuts; closures and redundancies. Templeborough is reduced to 4 furnaces.

### 1977 THE RECORD
Despite problems in the industry, 'E' Furnace famously breaks the world record for liquid steel output in a week – producing 74.2 tonnes an hour.

### 1979 STEEL STRIKE
To stem continuing losses, British Steel proposes more cut-backs. Fearing the social consequences, the unions call a strike. But to no avail. No more money is forthcoming. Nothing changes.

### 1981 THE PHOENIX PROJECT
A reduced industry is reborn. Fed by 'E' Furnace, the UK's first continuous casting machine is installed – capable of producing ½ million tonnes a year. Modernising continues until 1992. But it's not enough to save Templeborough.

## LEGACY FOR THE FUTURE
### 1990 SEEDS OF MAGNA
Rotherham Metropolitan Borough Council have the idea for an Iron and Steel Heritage Centre, to be located alongside the working Templeborough steelplant.

### 1993 THE LAST CAST
On November 25th 1993, 'E' Furnace makes its last ever cast. The plant is to close before the year end. Templeborough's spirit of innovation lives on in the modern steel industry.

### 1994 REVIEW AND REGENERATION
The Magna project is reviewed and developed over the next 5 years. It will now be in the Templeborough building. The name is registered. Plans and funding are put in place. The public consulted. Magna will be a world class visitor attraction and contribute to the area's economic regeneration.

### 1999 BUILDING STARTS
Preserving the original steelworks, while creating something completely new – the UK's first Science Adventure Centre. The work – including all 4 pavilions – has to be finished in just two years.

### 2001 MAGNA OPENS
On time. A new chapter in Templeborough's 2000 year old history begins...

# The future

**Magna is new. But it's growing. It's changing. All the time.** Future developments include flying robots, in formation. Predator and prey robots. Hunting. Running. Killing. Learning. Evolving. See them do it.

### MAGNA INVADED BY ROBOTS
You've seen Robot Wars, Techno Games and RoboCritters on TV. Now Magna is about to be invaded by robots. In conjunction with Professor Noel Sharkey and the Department of Computer Science at Sheffield University, it has set up CRUM – the Creative Robotics Unit at Magna – to explore the frontiers of state-of-the-art robotic research, and make it available and accessible for everyone to enjoy. Its first two projects are already underway and will be making their debut at Magna in the coming months. These are leading edge projects. Technologically advanced. Nothing virtual here.

### THE FLYBORG PROJECT. FLYING, FLOCKING ROBOTS
The 'Flyborgs' are an experimental group of robot blimps – flying balloons designed to fly independently and flock together in aerial displays, like birds or insects. Each Flyborg flies on its own, programmed to follow simple rules – like keeping its distance from the others. The idea being that gradually flocking behaviour will emerge. The Flyborgs will come together to create the flock – or 'Florg' – and fly in their allotted run, following the leader in their own form of formation flying. But remember – this is an experiment. We don't yet know exactly how it will work.

The Flyborg Project is unique. It has only been possible because of the huge indoor space available at Magna.

Magna won't stand still. It's a living, breathing, dynamic place. Already packed with amazing exhibits and spectacular shows, it will go on developing. Changing. Evolving. Adding new, innovative, ground-breaking, fun things to the amazing range of its current collection. There will always be more to find out. More to enjoy. Next up is robotics. The future. Coming soon.

### HOW THEY MIGHT DEVELOP

The Flyborgs are still in development. But they may carry on-board cameras to transmit live 'Flyborgs-eye' images, as well as being able to adjust to the unpredictable airflows in Magna – tilting and turning automatically, to fly more effectively. Later in the project, through user friendly computer interfaces, you may be able to control them. Altering the rules. Changing the movement of the Florg or the leader. We hope that, eventually, you will be able to fly the lead robot. In time, the Flyborgs may develop different behaviours. They may learn to adapt to new circumstances. There may even be two groups who communicate and interact. Creating even more spectacular displays.

'Autonomous Robotics' – is where robots operate without human intervention. Controlled by microcomputers. Taking input from sensors. Sending output instructions to motors.

'Biomimetic Robotics' – is where robots are used to test biological theories and animal behaviours. In this case, how complex behaviours can arise from simple actions.

### THE COMCO PROJECT. CO-OPERATIVE AND COMPETITIVE CO-EVOLUTION

If you think the Flyborgs are amazing, this is truly mind blowing. Imagine that by the year 2020, the priority for space exploration is finding new mineral wealth and energy resources. A special project sends two types of robot into space to lead the hunt. 'Search Robots' which seek out minerals and come back to scattered bases to recharge their batteries. And 'Collector Robots' that pick up weak or damaged 'Search Robots' and bring them back to a series of collection points, where their own batteries are recharged. In time, humans forget about them, and the robots begin to evolve. The 'Collector Robots' become predators. The 'Search Robots' their prey. This is the idea behind the Comco project. What happens next?

### THE FIGHT FOR LIFE

At Magna, there will be a group of prey robots and a much smaller group of predator robots. The prey will have solar cells on top of their bodies, and maintain their battery power by 'grazing' under 'light trees' distributed around the display arena. The predators will maintain their energy by catching the prey in their grippers and exchanging them for a battery charge. The prey robots will be quicker. The predators will be bigger. Both have the same senses.

### THE FITTEST SURVIVE

The robots are controlled by artificial brains (ANNs – artificial neural networks) and these can evolve – breeding by uploading their genes to a remote computer. Survival of the fittest will apply. Only robots that survive to 'maturity' will be allowed to breed. They'll only die by starvation. Predation. Or faults in the hardware.

### YOU'RE PART OF IT

Again, visitor information will be provided by computer. Again, the robots will work on their own. Or be controlled by you – through a user-friendly interface. Short-term, you'll see the hunt. Seriously scary. Long-term, you'll witness the process of evolution. Even scarier. Predators will get better at hunting with each new generation. They'll probably form hunting groups. Packs. Prey will get better at escaping. Individually, and by herding together in flocks. Co-operation and competition. Learning and evolution. Together. This has never been done before. And never over a prolonged timescale. This is a first. This is the future. Not happening in years to come. But in a few months time. At Magna.

### LOOK OUT FOR OTHER NEW DEVELOPMENTS AND SPECIAL EVENTS

Check regularly. Magna is always exploring new attractions and runs a constantly evolving programme of special events. If you'd like to know more about what's new and what's coming to Magna, just fill in a visitor information form at the Information Desk – and we'll keep you posted. Or visit our website at: www.magnatrust.org.uk. You never know what's around the corner. At a Science Adventure Centre near you. Magna. Expect the unexpected.

# In and around

What else do you want to explore? The Meadowhall Centre. Over 270 stores. Town. City. Country. Herons. Foxes. Newts. Bullrushes. Kestrels and weasels. Waterwheels. Hammers and steam engines. And there's always much, much, more.

### WHERE ELSE TO VISIT

Magna is in a great part of the country. It's minutes from Rotherham and Sheffield. Next door to the famous Meadowhall Centre. And thanks to the motorway and a good road network, it's only minutes more to Pennine Yorkshire and the Peak District. While you're here at Magna or when you're planning for subsequent visits, why not check it out. There's loads to see and do. Here's where you can shop up a storm. Discover a nature reserve where you'd least expect it. See and find out lots more about iron, steel and our industrial heritage. Go to town or enjoy the landscape.

### SHOPPING AT MEADOWHALL

Meadowhall is less than a mile from Magna. One of the biggest and best known out-of-town shopping centres in Britain, it attracts more than 30 million visitors a year. It has well over 12,000 parking spaces. A fantastic range of facilities. A crèche and play area. An 11-screen cinema. And an incredible 270 stores. Including more than 30 cafés, kiosks and restaurants. For more information, call the Meadowhall Careline on 0845 600 6800. Or visit the Meadowhall website: www.meadowhall.co.uk The average length of a Meadowhall shopping trip is 2 hrs and 7 minutes. Which leaves you plenty of time for Magna.

### BLACKBURN MEADOWS NATURE RESERVE
#### Unexpected Nature

This is another of Magna's close neighbours, and there are plans afoot to link the two by footbridge. Blackburn Meadows was once a farm on the River Don floodplain. Then for around 100 years, from 1886 onwards, it was a sewage farm, with huge liquid sewage lagoons. Surprisingly, these attracted all kinds of birds and wildlife. And when the sewage farm went, a beautiful and altogether unexpected nature reserve was created. See the Heron Gate. Dragonfly and Damsel Fly sculptures. There are wildflowers. Swans. Gulls. Birds of prey. Waders and rare migrating birds. It's open all year. Entrance is free. For more information, call Blackburn Meadows Trust on 01709 822041.

Magna is home to Sheffield Wildlife Trust's Education Centre, offering both classroom and hands-on educational activities in Blackburn Meadows. For more information, call 0114 2634335.

### A UNIQUE INDUSTRIAL HERITAGE
You can visit some terrific museums and heritage centres hereabouts. With some remarkable examples of water and steam power. Used for processing iron and steel, and grinding grain. Between them, they'll give you a unique insight into Yorkshire's industrial past.

### KELHAM ISLAND MUSEUM
Houses one of the world's most powerful steam engines. The mighty 12,000 horsepower River Don Engine. Built in Sheffield in 1905, it was used to make armour for Dreadnought battleships. Steel plates for North Sea oilrigs. Also see traditional Sheffield cutlery craftsmen at work. Call 0114 272 2106.

### ABBEYDALE INDUSTRIAL HAMLET
Has forged iron for over 500 years. Making scythes and hay knives since the 18th century. Waterwheels powered 23-ton hammers. Grinding wheels. Blowers. Provided power for the workshops. You can also explore the workers' and managers' houses. Call 0114 236 7731.

### SHEPHERD WHEEL
On the banks of the River Porter, is the only surviving water powered grinding wheel – once used to make all kinds of knives. To see it work, call Abbeydale on 0114 236 7731.
Also visit the Sheffield Museum Trust website at: www.simt.co.uk

### ELSECAR HERITAGE CENTRE
Has the world's only Newcomen Steam Engine still on its original site. It was used to pump water from local mines at 400 gallons a minute! Also at Elsecar, the PowerHouse Interactive Centre, a railway and craft workshops.
Call 01226 740 203.

### WORSBOROUGH MILL
In a 17th century mill, a wheel drives grinding stones for grain. In a 19th century mill, a rare Hornsby hot-bulb oil engine turns two big millstones. Call 01226 774 527.

### WORTLEY TOP FORGE
Is the only remaining ironworks of its kind in Britain. It used water power to make iron for nails and later, railway axles. The present, largely 18th century forge is complete with dam, sluices, hammers, cranes and 3 iron waterwheels. Call 0114 288 7576.

### TOWN AND COUNTRY
In the pages of the Magna Souvenir Guide, we simply don't have space to even begin to do justice to all that Rotherham, Sheffield and the surrounding countryside have to offer. What we can do is provide you with the briefest-of-brief introductions, and give you the telephone and fax numbers; e-mail and website addresses where you can find out more.

### PENNINE YORKSHIRE AND THE PEAK DISTRICT
The area offers beautiful, spectacular and often starkly dramatic landscapes. Attractive towns and villages. Markets. Crafts. Museums. Historic houses and heritage centres. And a whole range of country and outdoor activities. Walking, climbing, riding etc. If you haven't visited here, you should. If you have, then it's never too soon to come back.

### ROTHERHAM AND SHEFFIELD
Rotherham's a great town. It's 900 years old, with a cannon outside the Town Hall, a chapel on a bridge, a canal, street markets, and a huge country park, stately homes, pubs, restaurants and pretty villages just up the road. Sheffield is a vibrant city. It's built on seven hills, like Rome. It has great shops. A university campus. Famous rock venues, and the biggest theatre complex outside of the Capital. On the doorstep, along the motorway, are the Peak District National Park and Pennine Yorkshire. So when you've finished at Magna, you can take to the hills. To find out more, here's who to contact:

Rotherham Tourist Information Centre,
Central Library, Rotherham. Tel: 01709 835904
e-mail: TIC@rotherham.gov.uk
www.rotherham.gov.uk

Destination Sheffield, Visitor Information Centre, 1 Tudor Square, Sheffield S1 2LA.
Tel: 0114 221 1900
e-mail: info@destinationsheffield.org.uk
www.sheffieldcity.co.uk

Barnsley TIC, 46 Eldon Street, Barnsley S70 2JL.
Tel: 01226 206757. e-mail: Barnsley@ytbtic.co.uk
www.barnsley.gov.uk

Yorkshire Tourist Board, 312 Tadcaster Road,
York YO24 1GS. Tel: 01904 707070.
www.yorkshirevisitor.com

Other Millennium Projects to visit in the area: The Earth Centre (Doncaster), The Deep (Hull), The National Space Centre (Leicester). To find out about these and other Millennium Projects visit www.millennium.gov.uk

**Parties. Education. Study. Discovery. Dinners. Celebrations.** Doing business. Hospitality. Enjoyment. Information. The numbers. The addresses. The worldwide web. Who does what. Who to call. Who to speak to.

### CONTACT MAGNA

Magna's larger-than-lifeness, the space, the architecture, the steelworks, the exhibits, the combination of fun and adventure, design, art, science, technology and engineering – all combine to create a brilliant new resource in the region. For families. For children. Schools and colleges. Groups and clubs. Businesses and corporate entertainment. Find out about it. Use it. Enjoy it. There's nothing like it. Anywhere. Magna.

### CHILDREN'S BIRTHDAY PARTIES

Magna's a great place for children's birthday parties. Bookings include games and activities. Unlimited use of Magna and the outdoor adventure playground. A special party lunch or tea. Party packs and an optional birthday cake, if you want one. Brilliant.

### GROUPS AND CLUBS

Magna's a fun place to come to in a group. En masse. In a bunch. That's 12 people or more, who the staff at Magna will make very welcome. Benefits include: discounted admittance price; pre-booked entry times; free coach parking and entry for drivers. And special facilities and/or programmes by request.

### EDUCATION MAGNA

Magna is very much about redefining the relationships between art and science; creativity and industry; lifestyle and the environment. It offers a multi-sensory, creative and immersive experience. But at the same time addresses teachers' requirements to